Texas Applications and Problem Solving

Texas Applications and Problem Solving

For use with

Calculus

Eighth Edition

and

Calculus of a Single Variable

Eighth Edition

Larson/Hostetler/Edwards

Houghton Mifflin Company Boston New York

Publisher: Richard Stratton
Sponsoring Editor: Cathy Cantin
Development Manager: Maureen Ross
Development Editor: Yen Tieu
Supervising Editor: Karen Carter
Senior Project Editor: Patty Bergin
Art and Design Manager: Gary Crespo
Senior Marketing Manager: Jennifer Jones
Director of Manufacturing: Priscilla Manchester
Cover Design Manager: Anne S. Katzeff

Printed in the U.S.A.

ISBN 13: 978-0-618-75179-2
ISBN 10: 0-618-75179-3

23456789–CRS–10 09 08 07

Contents

Texas Applications

Chapter P Texas Applications and Problem Solving

Texas Gross State Product

The economy of Texas is often measured in Gross State Product. Gross State Product is an estimate of the total value of all goods and services produced in the state during a particular year. Increases in the Gross State Product of Texas reflects economic growth and development in the state.

The state makes forecasts concerning economic growth in order to plan its budget and anticipate what needs might occur in the state. According to a fall, 2003 forecast, this will be the trend for economic growth in Texas for the near future.

Years	Gross State Product (in billions of dollars)
2000	647.93
2001	689.53
2002	734.72
2003	782.53
2004	832.58
2005	886.79
2006	944.78
2007	1006.06
2008	1070.96
2009	1138.00
2010	1208.64
2011	1284.49
2012	1366.16
2013	1453.71
2014	1546.79

(*Source:* Texas Comptroller of Public Accounts)

In Exercises 1–4, use the table of data given above.

1. Graph the data above.
2. Which type of function best models this data: a linear, quadratic, or trigonometric function?
3. Use the regression function on a graphing calculator to find a model for this data. What does your model project will be the Gross State Product in the year 2020?
4. What does your model project for the year 2099? Do you think that this projection is reasonable? Explain why or why not.

Predicting Texas Population

The U.S. Census Bureau collects essential statistical information about the general population. Since this data is representative of the population over a finite period of time, mathematicians use this data to predict how the state population may change in the future.

State leaders use these predictions to plan for the necessities of the states; for example, what roads should be built, which services will be needed, and the amount of tax revenue that will be generated. The following table shows the projected population of the state of Texas, organized by age group.

Ages	2000	2010	2020
5–17	4,262,131	4,708,080	5,676,403
18–24	2,198,881	2,504,460	2,658,208
25–44	6,484,321	6,912,464	7,731,890
45–64	4,209,327	5,859,173	6,520,717
65+	2,072,532	2,587,383	3,755,814

(*Source:* U.S. Census Bureau)

In Exercises 1–6, use the table of data given above.

1. Use the regression function on your graphing utility to derive linear equations that model the trend in each age group. (Let $t = 0$ for 2000, $t = 10$ for 2010, etc.)
2. According to your projections, which age group shows the fastest growth?
3. According to your projections, which age group will represent the largest portion of the Texas population in 2050?
4. According to your projections, in what year will the population of individuals 65 years of age or older exceed that of those between 25 and 44 years of age?
5. What do your projections forecast as the populations for each of the age groups shown in the year 2075?
6. What do your projections show as time approaches infinity? Describe several factors that might change the trends you are projecting.

Chapter 2 Texas Applications and Problem Solving

Men Versus Women

Demographists and marketing professionals are among those interested in knowing the ratio of men to women. These statistics are used to track population changes, plan marketing strategies, and predict possible employment opportunities.

The United States population has experienced an overall decrease in the gap between the population of men and women. In 1990, there were 94.9 men for every 100 women, and in 2000, the ratio of men to women was 96.5 to 100. The following projections were published for the state of Texas.

Years	**2000**	**2010**	**2020**	**2030**
Population of Men (m)	10,350,000	12,220,000	14,200,000	16,510,000
Population of Women (w)	10,500,000	12,430,000	14,440,000	16,810,000

Source: United States Census Bureau

In Exercises 1–5, use the table of data given above.

1. Use the regression function on your graphing utility to find quadratic models of $m(t)$ and $w(t)$, where t is the time in years, with $t = 0$ corresponding to the year 2000.
2. Find dw/dt. What does this function represent?
3. Use the regression function on your graphing utility to find a linear model of $w(m)$. Use the derivative of this to find the value of dw/dt if it is predicted that the population of men in Texas will increase at the rate of 250,000 per year in 2010.
4. According to your projection, what is the population of women in the state of Texas when the male population is 20 million?
5. Predict the ratio of women to men in Texas in the year 2050. What other information would help you interpret the importance of this statistic?

Chapter 3 Texas Applications and Problem Solving

Spending by the Texas State Government

The Texas state government spends money on thousands of diverse budget items. To help keep track of this spending, the budget is divided into eight main categories. Education is the largest of these categories; the next largest is Health and Human Services. On the other hand, Regulatory and Judiciary costs are two of the smallest categories in the budget.

Many public groups track the spending on each of these categories from year to year. They are interested in making sure that the government does not waste money and that their particular group is getting its "fair share" of the money spent. The Texas Legislative Budget Board has released the following estimates for the years 2002 to 2007.

Categories	**2002–2003**	**2004–2005**	**2006–2007**
General	2663.4	2660.9	2960.5
Health	37,806.0	41,333.6	44,002.9
Education	49,661.9	63,140.8	54,704.4
Judiciary	428.8	442.6	444.6
Criminal	8410.8	8316.3	8156.8
Natural Resources	2190.7	2022.6	2323.3
Economic Development	13,900.7	14,321.0	15,638.2
Regulatory	717.6	842.2	687.6

(*Source:* Texas Legislative Budget Board)

In Exercises 1–4, use the table of data given above.

1. Use the regression function on your graphing utility to find and graph a linear model for each of the budget categories, where t is the time in years. (Let $t = 1$ for 2002–2003, $t = 3$ for 2004–2005, and so on.)
2. Does the projection for the Regulatory category of the budget suggest a time when this budget line will equal zero? If so, when will that be? Would it be reasonable to expect the budget trend to continue beyond that point?
3. According to your projections, which budget category will grow fastest? What are the projected results of this growth in 50 years?
4. According to your projections, what is the maximum predicted budget for Education? What is the maximum if a quadratic model is used instead?

Chapter 4 Texas Applications and Problem Solving

Wind Power in Texas

Texas uses wind power to generate electricity. A plan considered by the Texas Energy Planning Council in 2004 suggested that wind turbines in West Texas could provide a substantial amount of energy for the state. The cost of the wind power, measured in equivalents of million cubic feet of gas (Mcf), is projected to correspond with the following formula.

$$c = -\frac{t}{14} + 3, \text{ where } c = \text{cost and } t = 0 \text{ corresponds with the year 2000.}$$

(*Source:* Texas Energy Planning Council)

In Exercises 1–3, use the model given above.

1. Find the cost of one Mcf of wind power in the year 2007 and in the year 2010.
2. What is the cost (in million cubic feet (Mcf) of gas) of wind power for the period 2000 through 2010?
3. What does the formula predict for the cost of wind power in 2050? Is this prediction reasonable?

Texas State Debt

The Texas state government uses bonds to finance some of its spending. Each bond obligates the government to pay a fixed interest rate to the holder until the original amount borrowed (the principal) is paid back. One bond series issued in 2004 borrowed $14,249,398 at the equivalent of a continuous compound interest rate of 3.6175%.

The amount of interest can be found by the formula for continuous compound interest $A = Pe^{rt}$, where A is the amount of interest, P is the principal, r is the annual interest rate, and t is the number of years.

(*Source:* Texas Bond Review Board)

In Exercises 1–4, use the equation given above.

1. Graph the function of the state debt on this bond assuming that the state allows the debt to grow without making any payments for the next 100 years.
2. How much interest would the state owe on this debt at the end of the 100 years?
3. Assume that the state government pays $1,500,000 at the end of each year towards reducing the debt on this bond. The money is used to pay off the interest accrued for that year and then the remainder is applied to the principal. Approximately how many years will it take to pay off the bond under this assumption?
4. What percentage of the first year's payment will pay off the accrued interest? What percentage applies to the principal?

Chapter 6 Texas Applications and Problem Solving

Population of Texas

Every 10 years the U.S. Census Bureau conducts a census of the United States. There are numerous uses for this data, including determining the number of seats each state is entitled to have in the U.S. House of Representatives.

The following data shows the census population data every 10 years from 1950 to 2000:

Year	Population
1950	7,711,194
1960	9,579,677
1970	11,196,730
1980	14,229,191
1990	16,986,510
2000	20,851,820

(*Source:* U.S. Census Bureau)

In Exercises 1–4, use the table of data given above.

1. Use the 1950 and 1960 data to find an exponential model M_1 for the data. Let $t = 0$ represent 1950, where t is measured in decades.
2. Use a graphing utility to find an exponential model M_2 for the data. Let $t = 0$ represent 1950, where t is measured in decades.
3. Use a graphing utility to plot the data and graph both models M_1 and M_2 in the same viewing window. Compare the actual data with the predictions. Which model fits the data better? Explain.
4. The U.S. Census Bureau estimated that the population for the state of Texas would be 22,490,022 in 2004. How does this compare with the predictions from model M_1 and M_2?

Texas State Revenues

The Texas state government receives the majority of its revenue from sales tax. It also receives income from taxes on cars, gasoline, cigarettes, alcohol, motels and hotels, and oil and gas production. Other sources of income include the state lottery, sale or rental of state land, and grants from the federal government. Total revenues for Texas are shown in the following table.

Year	Total Texas Revenue
1998	44,497,247,142
1999	47,970,044,913
2000	49,845,829,550
2001	53,823,701,741
2002	55,221,546,458
2003	58,309,957,281
2004	62,073,072,643
2005	65,810,167,431

(*Source:* Texas Comptroller of Public Accounts)

In Exercises 1–3, use the table of data given above.

1. Use the regression function on your graphing utility to find an exponential model for the data. Let t represent the year, with $t = 1$ corresponding to 1998. Plot the data and graph the model in the same viewing window.
2. What does the model predict for state revenues for the year 2010?
3. A consultant believes that state revenues for 2005 through 2010 can be modeled by the function $60{,}000{,}000e^{\frac{2}{3}t} + 40{,}000{,}000{,}000$. What is the difference in total state revenues between the two models for the years 2005 through 2010?

Age of Texans

The 2000 census gave the following information about the age of people in Texas:

Age	Population
Under 5 years	1,624,628
5 to 9 years	1,654,184
10 to 14 years	1,631,192
15 to 19 years	1,636,232
20 to 24 years	1,539,404
25 to 34 years	3,162,083
35 to 44 years	3,322,238
45 to 54 years	2,611,137
55 to 59 years	896,521
60 to 64 years	701,669
65 to 74 years	1,142,608
75 to 84 years	691,984
85 years and over	237,940

In Exercises 1–4, use the data given above.

1. Plot the data using a histogram. Assume that the population in each age bracket is evenly split between years and that the "85 years and over" category is evenly split between the years 85 and 99.
2. Use a graphing utility to find a quadratic function that models the age distribution of Texans. Approximate the histogram by entering one data point for each five-year interval, using the ages 2, 7, 12, 17, etc.
3. Find the age at which a person would be older than two-thirds of all Texans. At which age would a person be younger than 88% of all Texans?
4. How do you think this Texas population distribution will compare to the distribution ten years from now? Name some factors that could affect the population distribution over the next ten years.

(*Source:* U.S. Census Bureau)

Median Family Income in Texas

Half of the families earn more than the median family income, and half earn less than it. Economists are interested in how the median family income in Texas changes over time. The following data from the U.S. Census Bureau shows the median family income for Texas in twelve different years. (Note: Data is in current dollars.)

1990	\$28,228
1991	\$27,733
1992	\$27,953
1993	\$28,727
1994	\$30,755
1995	\$32,039
1996	\$33,072
1997	\$35,075
1998	\$35,783
1999	\$38,688
2000	\$38,609
2001	\$40,860

In Exercises 1–3, use the table of data given above.

1. Use the regression capabilities of a graphing utility to find a model of the median family income of the form $a_n = b_n + c$, $n = 0, 1, 2, 3, \ldots, 11$. Graph both the data and the model.
2. What does your function project as median Texas family income for 2008?
3. Median family incomes in Texas for the years 2002–2004 were as follows: in 2002, \$40,149; in 2003, \$39,271; and in 2004, \$41,326. Use a graphing utility to find an updated model of the median family income of the form $a_n = b_n + c$, $n = 0, 1, 2, 3, \ldots, 14$. Project the median family income for 2008 with the new model. What impact, if any, do these three additional points have on your projection for median Texas family income for 2008?

(*Source:* U.S. Census Bureau)

International Space Station Sightings

The International Space Station is a multi-national project. Its first component was launched in 1998, and the station was first occupied in 2000. It is 240 feet wide, weighs over 400,000 pounds, and has 15,000 cubic feet of habitable area.

For the week from January 26, 2006 to February 1, 2006, the International Space Station was scheduled to be visible in its orbit over Amarillo, Texas, six times as follows:

Date and Time		Duration	Max. Elev.	Approach	Departure
Jan. 26	07:11 a.m.	2 min.	31°	10° above SSW	31° above SSE
Jan. 28	06:25 a.m.	2 min.	38°	11° above SSW	38° above SE
Jan. 29	06:50 a.m.	4 min.	44°	19° above WSW	11° above NE
Jan. 30	07:15 a.m.	4 min.	15°	10° above WNW	10° above N
Jan. 31	06:07 a.m.	1 min.	21°	21° above NNE	11° above NNE
Feb. 1	06:31 a.m.	1 min.	13°	13° above N	10° above N

(*Source:* National Aeronautics and Space Administration)

A geographical direction such as SSW indicates a direction exactly halfway between South and Southwest. In order to describe the location of the Space Station using equations and a graph, we can replace these directions with their numerical equivalents; replace N with 0°, and proceed clockwise making East 90° and so on. Accordingly, SSW becomes 135°.

Assume that when the space station is visible on January 30 it moves uniformly from WNW to N along the arc of a circle and that it reaches its maximum elevation halfway between its approach and departure.

In Exercises 1–5, use the information given above.

1. For January 30, find the numerical equivalent of each compass direction as well as the numerical direction of the Space Station at its highest point. Use 360° for North.
2. Plot these three given points on a coordinate plane with numerical direction on the horizontal axis and elevation on the vertical axis. Sketch the path of the Space Station between its arrival and departure.
3. Find an equation in x and y that models the observed path of the Space Station on January 30.
4. Find the parametric equations that model the observed motion of the Space Station on the same day, letting $t =$ time in minutes.
5. What was the elevation and numerical direction of the Space Station at 7:18 a.m. that morning?

Spherical Coordinates in Texas

Earth is approximately a sphere with a diameter of 8000 miles. This makes it appropriate to use spherical coordinates as a way of describing the location of various places on the Earth's surface. It is similar to the latitude-longitude system when identifying cities. The latitude and longitude of some of the cities in Texas are given in the following chart.

City	°North	°West
Austin	30.32	97.77
Corpus Christi	27.68	97.29
Houston	29.77	95.39
Jefferson	32.77	94.35
Lubbock	33.59	101.89

In Exercises 1–4, use the information given above.

1. Find the spherical coordinates for the location of each city.
2. Find the rectangular coordinates for the location of each city.
3. Find the angle in radians between the vector from the center of the Earth to Lubbock and the vector from the center of the Earth to Corpus Christi.

   Use the formula $\cos\theta = \dfrac{\mathbf{u}\cdot\mathbf{v}}{\|\mathbf{u}\|\,\|\mathbf{v}\|}$.
4. Find the great-circle distance s between the cities.

   Use the formula $s = r\theta$:

Football in the Astrodome

Two football teams are playing in the Houston Astrodome. A kicker on one of the teams is standing directly in front of, and 30 yards from, the goalpost of the opposing team. He kicks the ball from the ground at an angle of 42° with respect to the ground through the goalpost to make a field goal. (The area through which he must kick the ball is at least 10 feet high and 18 feet 6 inches wide.)

In Exercises 1–3, use the information given above.

1. With what initial velocity must a football be kicked to get the football through the center of the goalpost? (Note: Assume a height of 10 feet is sufficient.)
2. What is the maximum height of the football after it is kicked?
3. How long does it take the football to reach the goalpost?

Chapter 13 Texas Applications and Problem Solving

Education Spending in Texas

The Texas state government categorizes most of its spending on education into three areas: Foundation School Program Grants, Higher Education Grants, and Other Public Education Grants.

Education Expenditures in Billions of Dollars

	Foundation	Higher Education	Other
2000	10.8	0.9	2.5
2001	11.2	0.9	3.0
2002	10.3	1.0	3.2
2003	10.6	1.0	4.0
2004	10.1	1.0	3.9
2005	10.6	0.9	4.0

(*Source:* Texas Comptroller of Public Accounts)

Mathematical models of total Texas State Government Education spending as a function of x, y, and z are based on this data. $x = 11 - 0.3t + 0.03t^2$ represents Foundation spending, $y = 0.8 + 0.09t - 0.02t^2$ represents Higher Education spending, and $z = 2.6 + 0.6t - 0.06t^2$ represents Other spending with $t = 0$ in the year 2000.

In Exercises 1–6, use the table of data and mathematical models given above.

1. Use the models to find the amounts of spending in each area and then calculate the product of the absolute values of x, y, and z, (cubic dollar space) for the period 2000 through 2010.
2. In which calendar year does the cubic dollar space of the three categories of spending reach its maximum? What is that amount?
3. In which calendar year will the cubic dollar space reach a minimum? What is the minimum?
4. Which of the three areas does your analysis suggest will grow the most during the period from 2005 through 2010?
5. Add the three functions together. What does the new function suggest will happen to total education funding by the state of Texas from 2005 through 2010?
6. An analyst suggests that state funding for Other education programs determines the amount that the state spends on Foundation and Higher Education. Calculate Foundation and Higher Education as functions of Other. (Graph Foundation and Higher Education as two separate lines on the y-axis with Other on the x-axis.) Does this function look reasonable? Explain.

Dome Houses

A company in Italy, Texas, designs and builds unusual structures that they call Monolithic Domes, structures which are spherical or oblate in shape. Customers can decide the diameter, height, and other dimensions of these buildings. One possible choice for a building is a portion of a hemisphere, 36 feet in diameter, on top of a cylinder with a diameter of 18 feet.

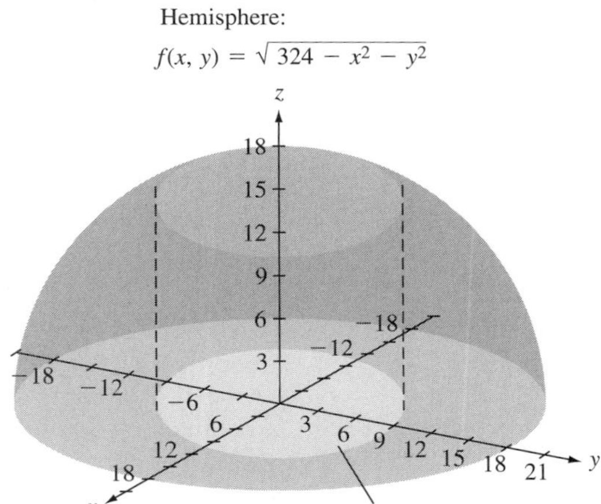

In Exercises 1–3, use the information given above.

1. Find the equations for the sphere and cylinder using cylindrical coordinates. Assume that the floor is at ground level.
2. Use integrals to find the volume of the structure. (Hint: Use the equation of a hemisphere whose center is at the origin. This hemisphere intersects a cylinder centered on the origin.)
3. Use integrals to find the surface area of the structure.

Airplane Navigation

A small plane was flying southeast from Odessa to Corpus Christi, a distance of approximately 418 miles. The plane maintains an airspeed of 125 miles per hour, and during the trip it encounters wind blowing precisely from the southwest. The wind's speed increases linearly with distance from 41 miles per hour over Odessa to 59 miles per hour over Corpus Christi. Assume that the pilot adjusts the plane's bearing in order to fly in a straight line from Odessa to Corpus Christi.

In Exercises 1–4, use the information given above.

1. Find the formula for the ground speed of the plane along the flight path in terms of distance x traveled.
2. Calculate the speed of the plane at distances 0, 209, and 418 miles.
3. How much time will it take the plane to travel from Odessa to Corpus Christi? (Hint: Since distance is found from $\int (speed)dt$, the time can be found from $\int \frac{dx}{(speed)}$, where t is time and x is distance.)
4. How much time would it have taken the plane to cover the same distance with no wind? Assuming that fuel use varies directly with time, what percentage of fuel was used to compensate for the wind?